The Geology of Montana De Oro State Park:
An uplifted marine terrace domain

& Morro Bay State Park
The last peak of a volcanic arc system

By William A. Szary

Copyright 2019. Earth2Energy. All Rights Reserved.

Book Cover: Wave cut terrace exposed at Corallina Cove in Montana de Oro State Park (front cover). Back cover is a view facing west of Morro Bay estuary and the pedogenic soil deposited on top of the surface.

Library of Congress Catalog in Publications Data:

Szary, William A.

Includes references

ISBN 13: 9781794168053

Key words: Montana de Oro, Morro Bay, central coastal California geology, marine terraces, volcanic arc system, Morro Rock, pedogenic soils, marine estuary

Earth2Energy Educational Publishing
Port Richey FL 34668

Earth2Energy is a Registered Trademark

Contents

Preface

Part I. Montana de Oro State Park

> *Introduction 5*
> *Geologic History 5*
> *Paleogeography 8*
> *Geomorphology 10*
> *Stratigraphy 12*
> *Structural Geology 14*
> *Park Geology Discussion 19*
> *Walking Tour Guide 21*

Part II. Geology of Morro Bay State Park 28
> *Geologic setting 28*

References 34

Preface

Montana de Oro State Park bedrock has been mapped several times by Pacific Gas and Electric Company, by the US Geological Survey in conjunction with the California Geological Survey in 1979 for the Nuclear Regulatory Commission included as part of a geological map compiled for San Luis Obispo County at a scale of 1:49000. Geologic mapping at this scale requires inferences of lithologic contacts over a large region from few exposure observations. Mapping at this scale ignores the occurrence of structural features easily overlooked due to the impossibility of examining every outcrop over the area covered by small scale mapping techniques, although geologic maps can be reasonably accurate at these small scales. Time and budget constraints also prevent larger scale mapping of areas.

The following presentation introduces a geologic map of Montana de Oro State Park at a scale of 1:24,000 presented by the California Geological Survey. Mapping at this scale enlarges the structural features for better understanding of the structural history of any given area. The geologic map presented in Figure 9 of this study is not intended to preclude any previous work conducted by other entities. Figure 9 shows the same structural folding patterns magnified 2x times. This book is intended for those unfamiliar with the geology of Montana de Oro State Park. It is written for those whom are familiar and unfamiliar with geologic terminology.

Special acknowledgements are provided to the Montana de Oro State Park staff, and to the Morro Bay Association for their help in providing contact support. Special thanks are given to Dr. Bauer, Professor at Cuesta College for helpful conversations pertaining to the geological history of the area.

Part I. Montana de Oro State Park

Introduction

Montana de Oro State Park is located south of Los Osos, California in San Luis Obispo County. The park is bounded to the north and east by the Los Osos State Preserve and to the south by PG&E Co. owned land.

Geologic History

To appreciate the geology of Montana de Oro State Park, a brief summary of California geologic history is presented here. Currently accepted theories propose that the earth is separated into approximately twelve crustal plates which slide over semi plastic like material called the mantle. **Figure 1** shows the approximate locations of these plate margins.

Figure 1. World plate tectonic map showing the mid oceanic ridge (center of Atlantic Basin) and various subduction zone complexes. Source: National Geographic Society.

The plate boundary type shown in **Figure 2a** represents the spreading ridge system type of plate margin. The Atlantic Basin represents this kind of plate motion where fresh molten lava is erupted from a central vent onto the sea floor. Over time, layers of lava, mostly of basalt composition build up and push away from the source.

Another kind of zone encountered in plate tectonics belongs to an obduction zone where oceanic crust is pushed upwards on top of continental crust when two plates collide. This most likely occurs when two continental plates collide against each other. This is why marine rocks pile up above continental rocks in mountain belt zones (**Figure 2b**).

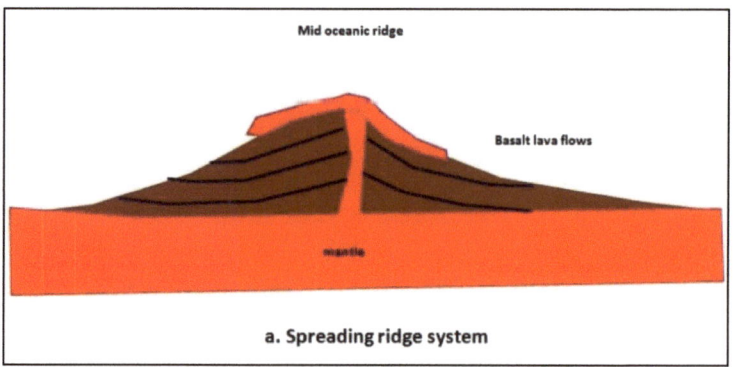

Figure 2a. A spreading ridge center develops when a central mid oceanic ridge taps into the upper mantle and allows molten lava to reach the sea floor. The lava begins to spill out, creating new oceanic crust of basaltic composition. As it builds vertically, it pushes outward away from the central vent to become new ocean floor.

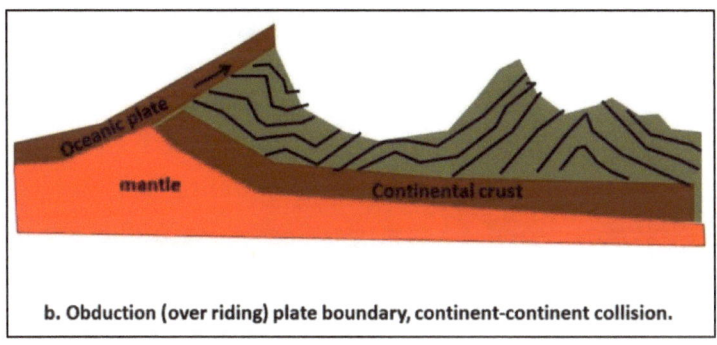

Figure 2b. Another type of plate boundary includes an obduction zone where two continental plates collide leaving oceanic crust on top of continental crust.

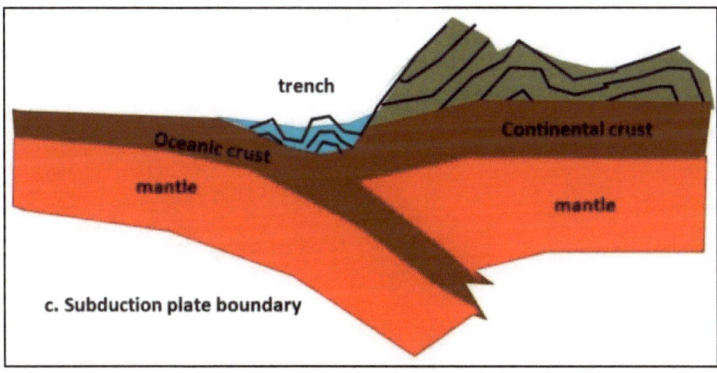

Figure 2c. A subduction zone type boundary occurs when the oceanic plate dives down into the mantle along a continental margin. A trench forms on the ocean floor above the zone which contains marine sediments and upper mantle rock that is deformed by the friction occurring between the oceanic and continental plate margins.

The San Andreas Fault zone began as a normal subduction plate margin (**Figure 2c**), forming the Coast Ranges approximately 150 million years ago (Middle Mesozoic Period) as the Pacific Plate collided with the North American Plate and was forced beneath it. Erosion lowered the Coast Ranges and deposited sediments into the trench formed by the downward travelling Pacific Plate.

Those sediments were then accreted (scraped off) onto the North American Plate adding land mass to what is now referred to as California (Bateman, 1974).

Following accretion of the sediments, currently accepted theories proposed that the Pacific Plate mid oceanic ridge (Figure 5a, 5c) migrated to the North American plate margin and separated at the mid oceanic ridge axis, thus forming the San Andreas Fault. The Pacific Plate changed direction from an eastwardly movement to a northwesterly movement, providing the slippage now currently associated with movements along the San Andreas Fault zone (**Figure 2d**).

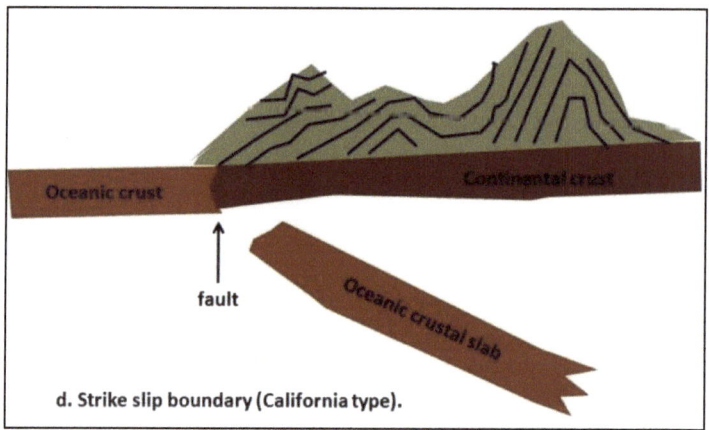

Figure 2d. A strike slip boundary or transform type boundary occurs when the subduction slab breaks off from the main oceanic slab and becomes directly in contact with continental plate margins. A shift in plate motion usually triggers formation of a transform type boundary.

Paleogeography

The Pismo Formation is approximately Pliocene to Miocene in age. The formation was deposited across the time line boundary. **Figures 3 and 4** provides a graphical depiction of conditions present along the western US coastline during this time. The coastal area between Morro Bay and Santa Barbara was submerged at this time.

Submergence allowed silts and clays to accumulate in moderate to deep water marine conditions.

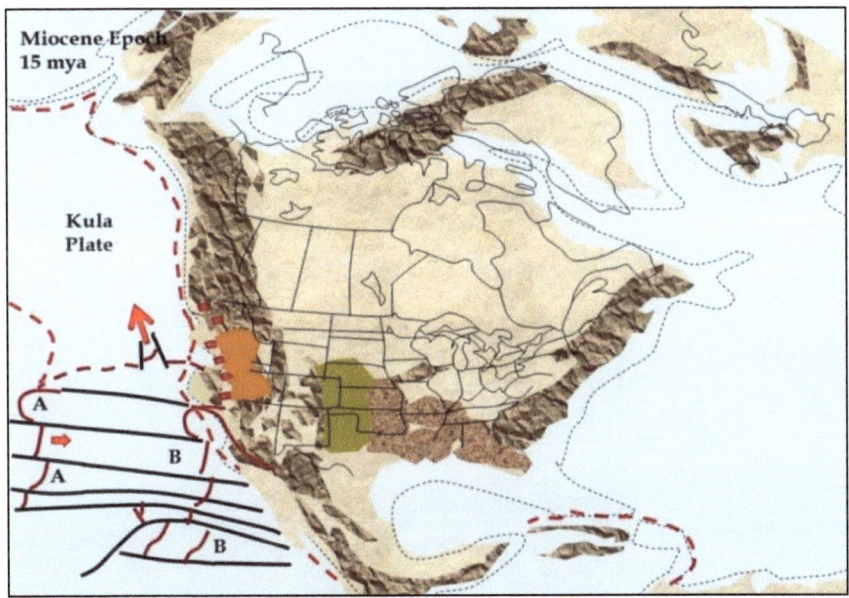

Figure 3. Paleogeographic conditions present along the western US at the end of Miocene time. Montana de Oro State Park was submerged at the time. Source: Szary, 2014.

Figure 4. Paleogeographic conditions present at the beginning of Pliocene time show the state park continued to remain submerged. Source: Szary, 2014.

Geomorphology

The park is located in the Coast Range geomorphic province as shown in **Figure 5**.

Figure 5. Geomorphic Provinces of California. Source: Hummert, 1978.

Montana de Oro State Park is part of the southernmost extension of the Santa Lucia Range and has two distinct sections which exhibit distinctly different topographical characteristics. The area north of Islay Creek consists of gently rolling to moderately steep sloped hills whereas the area south of Islay Creek consists of steeply sloped hills.

Elevations differ in both areas of the park as shown in **Figure 6** (published by the California Dept. of Parks and Recreation). North of Islay Creek, the highest elevation is Hazard Peak (1076 feet) which is approximately 300 feet lower than the highest elevation in the area south of Islay Creek (Oats Peak-1373 feet).

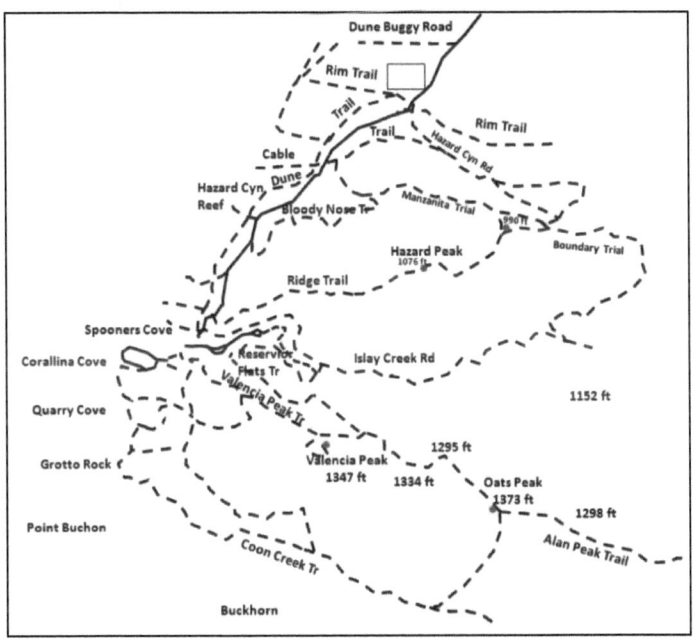

Figure 6. Montana de Oro State Park detail map. Source: California Department of Parks & Recreation.

Drainage patterns are similar throughout the park taking on a dendritic pattern (**Figure 7**). South of Valencia Peak, drainage patterns take on a parallel pattern particularly on the south side of Valencia Peak and on the north slope of Buckhorn and Last Peaks. Outstanding landforms occur in the park and may be observed along the coastline, south of Spooner's Cove. As one walks along the bluff and looks back east at the park hills, two ancient wave cut terraces can be viewed. One terrace is the bluff itself, appearing as a platform like bench. The terrace extends to east of Pecho Valley Road up to the first moderately steep slope on the east side of the road.

The second wave cut terrace is expressed in the hills which lie in the foreground of Hazard and Valencia Peaks.

The higher elevation peaks such as Valencia Peak and Oats Peak are called flat iron landforms formed as a result of erosion of tilted bedrock (named after irons used to press clothes).

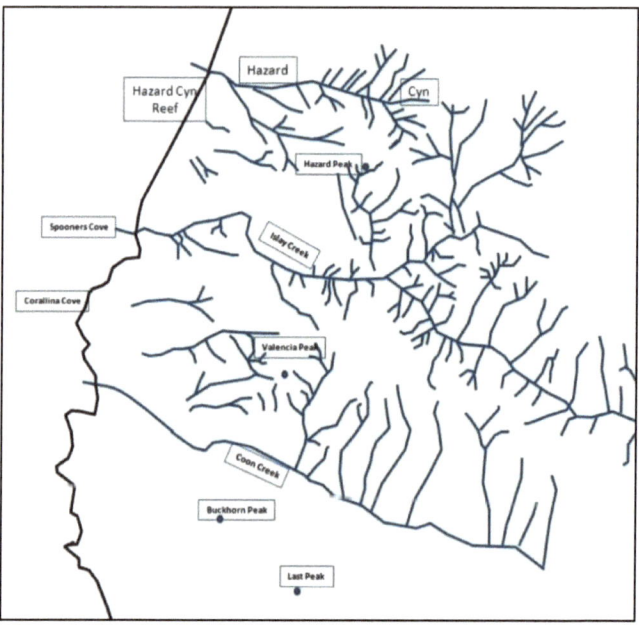

Figure 7. Drainage network in Montana de Oro State Park used to estimate rock types. Source: California Department of Parks & Recreation.

Stratigraphy

The Miguelito Member occupies the majority of land area within the park. It consists of buff to light brown siltstone, shale, porcellaneous shale, opaline shale, diatomaceous shale, minor friable sandstone that is well bedded and relatively soft. The shales encountered in the park were siliceous and cherty mudstones. Mudstones consisted of very fine silt size grains difficult to distinguish through a 10x hand lens. Fresh surfaces vary in color from shades of gray (brownish, greenish, whitish, and chalky). The mud rock weathers to yellow red and brown red iron oxide stain. Black staining (manganese) may also be present.

No fossils were observed. The Pismo Formation is shown in light brown on the geologic map. Cherty mudstones are distinguished from siliceous mudstones by the presence of laminations. Laminae are individual layers not greater than 0.5 cm (0.25 inches) thick.

Laminae alternate in colors between medium brown, brown gray, and chalky gray. Two other rock types are present. Sandstone and siltstone breccias overlie the Pismo Formation with angular unconformity. An unconformity represents a period of erosion before deposition resumes again (**Figure 8**).

Figure 8. Angular unconformity of sediments resting above the Pismo Formation Miguelito Member shale along the coastline in Montana de Oro State Park. The exposure is located at Spooner's Cove.

The unconformity may be observed along the sea cliffs at Spooner's Cove. Figure 8 depicts an angular unconformity in cross sectional view. Both lithologies can be separately identified from each other by differences in the grain size which make up the formational units. Siltstone breccias (upper part) consists of 80% fine silt sized grains difficult to distinguish with a 10x hand lens.

Breccias consist of 15% very angular mudstone fragments, are very friable when wet and are calcite cemented. Grains within the rock are poorly sorted. Sandstone breccias consist of medium to coarse grained sand which is 75% quartz (gray minerals). Sand grains are sub-rounded to sub angular. Sand breccias consist of 20% angular mudstone fragments with 5% consisting of plagioclase grains (white minerals). Rocks are very friable when wet and are moderately sorted. There also occurs a sandstone bed similar in composition to the sandstone breccia but is devoid of mudstone fragments. Sandstone is easily viewed with a 10x hand lens. Area geologists consider these brecciate units as ancient wave cut terrace deposits. The presence of angular mudstone fragments and the sub-angular to sub-rounded silt and sand grain shapes suggest their origin was continentally derived, probably from a source nearby possibly from the Pismo Formation itself. Rounded grains would imply long distance transport from a distant source.

Structural Geology

Figure 9 is a geologic map produced by the California Geological Survey. Two major folds occur in the park and are represented in the geologic cross section presented in **Figure 10**. Hazard Canyon is the axis of an anticline. If one proceeds to walk along Hazard Canyon Road, the north and south canyon walls are tilted opposite to each other. For every anticline, there is an associated syncline. The adjacent synclinal axis belongs to Islay Creek canyon. Valencia Peak and the area south of the park are the beginning of another anticline. The geologic map indicates there are disruptions in the axial trends, particularly noticeable in the Islay Creek syncline. Secondly, there occurs minor folding contained within the major folds. These folds are referred to as flexural and drag folds and are presented in **Figure 11**.

Flexural folds may be observed along the south wall of Coon Creek canyon. Drag folds are present opposite the rear campground on Islay Creek Road at the Reservoir Flats trail head. All major folds plunge. Plunging is defined as an end of a fold that is tilted in a given direction. Evidence of offsetting axes, flexural, and drag folds support the presence of major and minor faults.

Minor folding may be observed along the sea cliffs south of Spooner's Cove. Similar anticlinal axes occur north of Grotto Rock, in Corallina Cove, and north of Spooner's Cove. Smaller synclinal axes occur adjacent to anticlines and are exposed north of Grotto Rock, in Quarry Cove, and in Corallina Cove (**Figures 12 & 13)**. Beds of the Pismo Formation are variable in orientation in two distinct areas of the park. The first area is located near the old barn on Islay Creek Road and the second area is located along Islay Creek Road northeast of the campground.

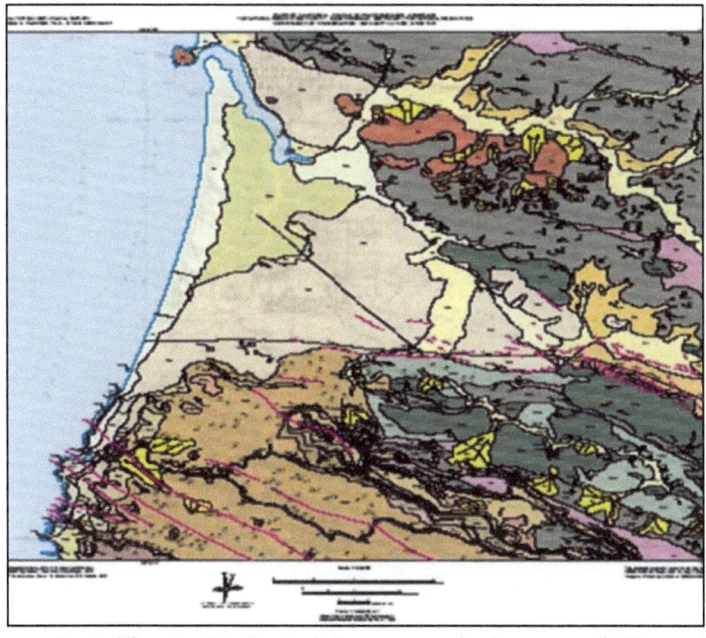

Figure 9. California Geological Survey geologic map of Morro Bay (northernmost portion of map) and Montana de Oro State Park (southern half of map). Red lines represent fold axes in the park. Source: Wieger, California Geological Survey, 1979.

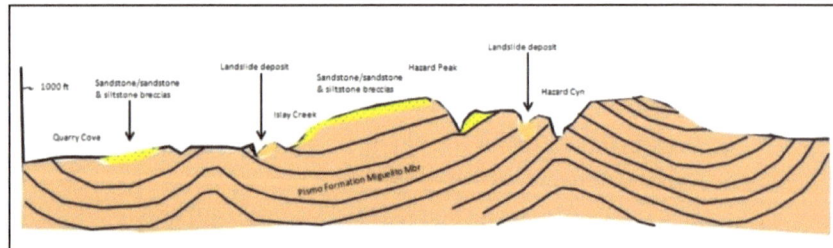

Figure 10. North to south cross section showing marine terrace deposits and landslide deposits overlying the Pismo Formation Miguelito Member.

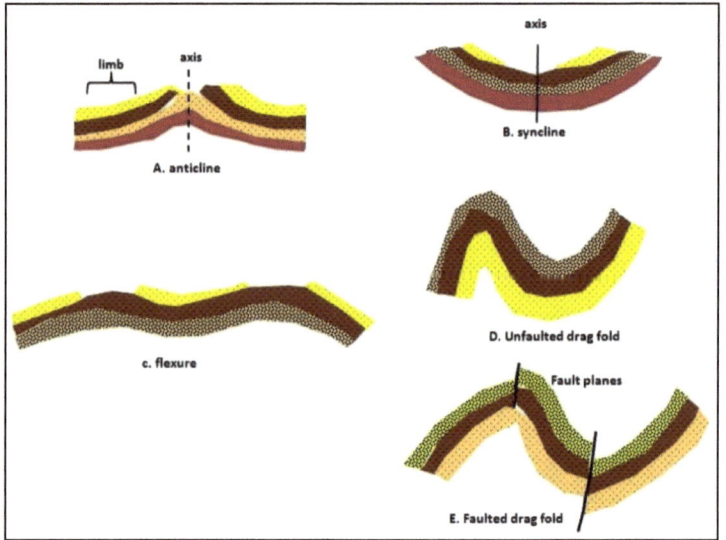

Figure 11. Major and minor folds encountered in Montana de Oro State Park.

Three possible causes are proposed to help explain these phenomenon: 1) landslide block slumping; 2) major or minor faulting associated with the currently accepted theory of multiple uplifts; 3) minor faulting occurring within the major folds. Evidence exists for support of all three proposals. Land-sliding is common throughout the park but is especially noticeable in the Hazard Peak area. The dominant rock type is sandstone and siltstone breccias. These rocks are very poorly cemented by calcite and crumble easily when moist.

Combine this characteristic with the criteria that these sediments overlie tilted mudstones. The tilting will act as a plane for both breccias and mudstones to slide down.

Figure 12. Pismo shale exposures along the coastline overlain by marine sediments in an angular unconformable relationship. The lower shale exhibits a synclinal fold. The black dashed line represents the axis. The corresponding anticlinal fold is located to the right of the photo, the limb of which is exposed in the upper portion of the photo.

Pismo mudstones, during the process of uplift, became distorted through minor fold development to the point where the mudstones are fractured along the bedding planes and, in many cases, parallel to the minor fold axes. Weathering of the mudstones is accelerated, allowing separation of large blocks which slide down the steeper slopes. Major and minor faulting may also help explain the variable orientations of mudstone. Minor anticlinal, synclinal, flexural, and drag folds are commonly associated with faulting. Fault planes are difficult to distinguish in outcrop. An example of a fault is shown in **Figure 14**.

Figure 13. A syncline-anticlinal sequence is exposed in this photo. The syncline is located to the right, in the beach area. The anticline is located to the left (white dashed line).

Figure 14. Displacement of rocks due to faulting may be observed along the Valencia Peak trail on the way to the peak. Rock orientation provides the clue. The upper slope at the top of the photo is oriented north to south. The floor of the trail is oriented northeast to southwest leaving the fault plane at the base of the upper exposure.

Minor folds are common to the mudstones. Flexural folds may be observed in outcrop locations 9 and 30 (**Figure 15**). Drag folds may be observed in outcrop location 111. Distortion of the major fold limbs may be an expression of minor folds.

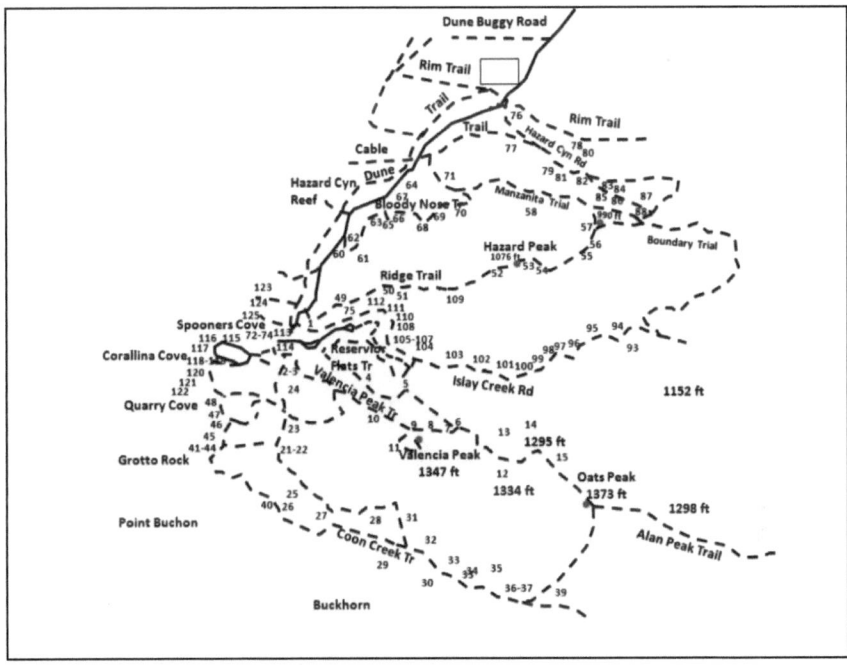

Figure 15. Montana de Oro State Park trail map showing locations of outcrops visited throughout the park. Source: California Department of Parks and Recreation.

Park Geology Discussion

The Pismo Formation was formed from silts and clays deposited in offshore marine waters. The presence of siliceous cementation, chert, silts, and clays suggest deposition occurred below the depths which permit calcite to exist, called the carbonate compensation depth. The gray color of the mud stone suggests a slight deficiency in oxygen. Fine grained sediments are commonly associated with calm water deposition not influenced by wave action.

Figures 3 and 4 show the approximate location of the shoreline as it existed along the California coast during Miocene-Pliocene times. The absence of fossils suggests deep water deposition of the mudstones occurred.

Fossils are abundant in rocks representing shallow to moderate depths which allow calcite to exist. Area geologists propose multiple uplifting of the Pismo Formation occurred after Miocene time (Chipping, verbal comm.).

Supporting evidence may be observed from the bluff as one looks back east into the park hills. Two distinct wave cut terraces are apparent: 1) the platform like terrace that extends up to Pecho Valley Road, and 2) the platform like terrace which makes up the lower elevation hills east of Pecho Valley Road. Both of these terraces were once located at sea level and were uplifted to their present position (**Figure 16**). In addition, one might observe from the bluff tilted beds of the Pismo Formation in the lower park hills (**Figure 17**).

This condition suggests different stages of uplift increased the tilt angle of the back hills during tilting of the fore hills. One stage of tilting may have occurred when the volcanoes represented by the volcanic necks along Highway 1. **Figure 18** helps explain the process which took place during each stage of uplift.

Walking Tour Guide

A walking tour guide of Valencia Peak trail was posted on the web without citing the author. It is reproduced here with modified drawings attached. **Figure 19** presents a detail map of the stops, terraces, trails, and creeks present in the park. **Figure 20** presents a profile showing the terraces along Valencia Peak trail as one would encounter them on a walk.

Figure 16. Marine terraces dominate the landscape throughout the park. The numbers provide the terrace locations beginning with No. 0 which is the marine cut platform currently being formed from current sea levels. No. 4 is the oldest uplifted terrace.

Figure 17. Tilted beds to the right represent the limb of an anticline fold to the left and a synclinal fold to the right. The trend persists in the lower hills. View is from Valencia Peak facing west.

Stop No. 1. – Spooner's Cove beach. The rock exposed in the lower half of the cliff is shale. About 6 million years ago (Pliocene Epoch) an ocean occupied the coastline and extended many miles eastward. This is the Miguelito Shale member of the Pismo Formation. About 3 million years ago, the shale was folded. Flat lying gravels cover the tilted shale in angular unconformity. The gravels were deposited about 100,000 years ago (**Figure 21 & 22**).

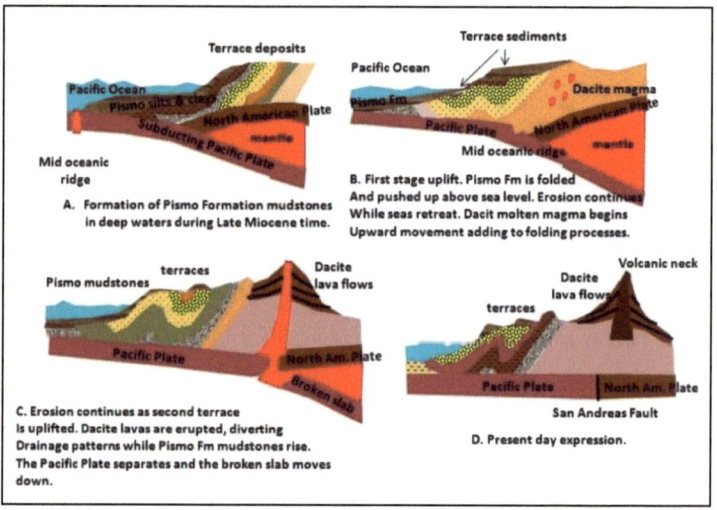

Figure 18. Process by which the set of marine terraces uplifted during Late Miocene into Pliocene time in the park.

Stop No. 2. Terrace 1. The first terrace is underlain by folded shale with gravel covering the surface of the eroded edges on top of the shale. During the past 100,000 years, marine erosion removed the upper shale that connected the sea stack with the cliff at stop No. 1.

Stop No. 3. Under the power pole line, facing the ocean. When the terrace #1 was at sea level, 100,000 years ago, the gravel was deposited as a wave cut terrace. The slope climbed onto for Stop No. 3 belongs to the eroded remnant of a vertical cliff cut by the ocean when Terrace #1 was being eroded.

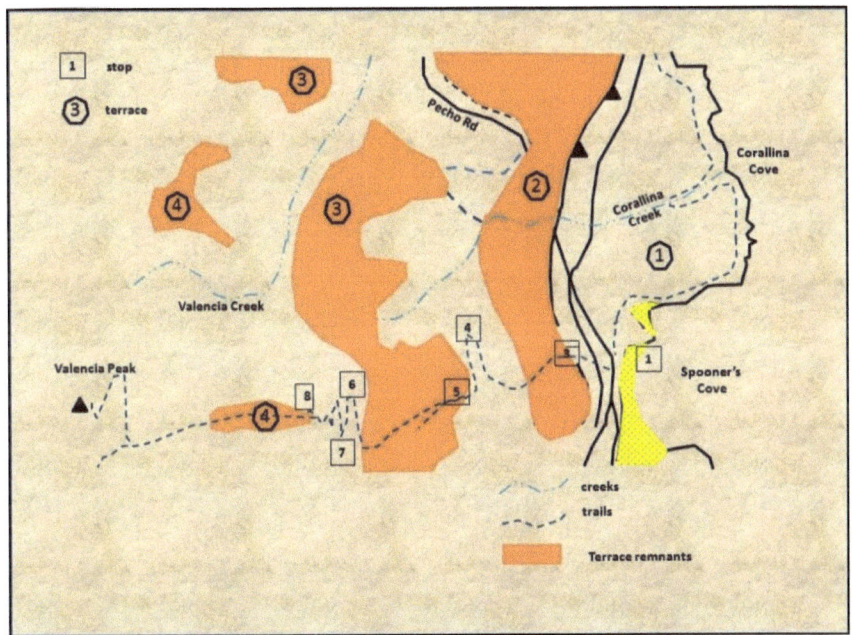

Figure 19. Map showing the terraces and stops encountered on a walking tour along Valencia Peak trail. Source: uncited publication posted on the web.

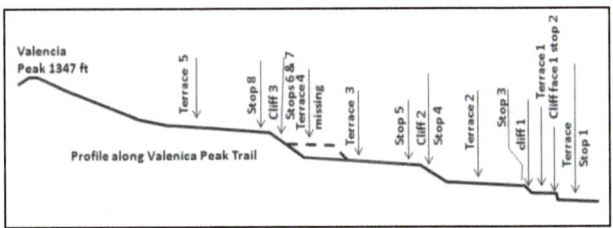

Figure 20. Profile along the Valencia Peak trail showing the terraces encountered.

When the terrace was uplifted, the ocean retreated westward about ¼ mile beyond the present day shoreline and started eroding a new terrace which is slightly below sea level. Terrace #0 has been eroding for 100,000 years.

Proceeding on the trail at the junction with Rattlesnake Flats trail, terrace #2 is being crossed. Terrace #2 was at sea level 150,000 years ago. In front is a low hill which belongs to an eroded remnant of the cliff that stood inland of Terrace #2 when it was located at sea level.

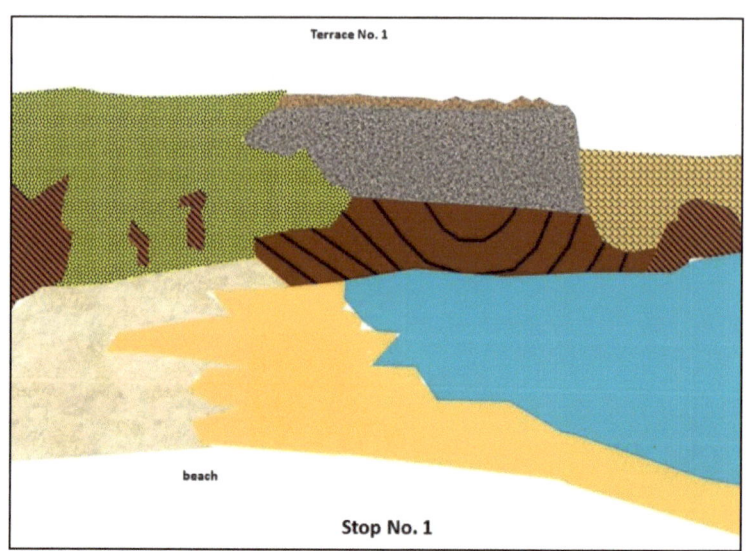

Figure 21. Stop #1 at Spooner's Cove represents terrace no. 1. The Miguelito Shale is the deformed brown unit beneath the gravel.

Stop No. 4. A short distance beyond the switchback is a large outcrop of Miguelito Shale at the top of the cliff behind Terrace No. 2.

Figure 22. Angular unconformity exposed along the southern wall of the coastline trail in the park.

Facing the ocean, Terraces 1 and 2 can be viewed. Terrace 3 is behind you about 300 feet inland. On the right is an outcrop of gray and orange sandstone. Accumulating sand was cemented beneath terrace 3. The orange represents iron rich deposits which leached from the gray areas.

About 150 feet further, an outcrop of gravel lies on the right side of the trail. This gravel belongs to stream deposits draining the hillsides when terrace 3 was at sea level. It rests on the eroded edges of Miguelito Shale. The geologic contact between the units is below the surface. About 100 feet further along the trail, you are walking on gray and orange sandstone.

Stop No. 5. You are positioned on top of terrace 3. Corallina Creek is eroded into terrace 3 when it uplifted 200,000 years ago. North of the trail, the same gray to orange sandstone is exposed in the gully. Facing eastward along the trail, groundwater feeds the gully servicing Poison Oak Spring and Railing Spring perched on top of the well cemented sandstone.

At the base of the eroded cliff of Terrace 3 the trail enters a junction. The left trail leads back to the Visitor Center. The right trail is Badger Trail (no sign) leading to the walk in campsite and the middle trail leads to Valencia Peak. The first switchback is Stop No. 6.

Stop No. 6. Coon Creek lies to the south. Point Buchon is south of Coon Creek. The canyon immediately south of this stop is being cut by Corallina Creek. The canyon is deeper through terrace 3 than it is through terrace 2 and 1. Terrace 3 was eroded for a longer time than the lower terraces. Each succeeding terrace from the ocean landward is older, subjected to longer periods of erosion. Terrace no. 1 is almost uninterrupted in length along the ocean front whereas only patches of terraces 2 and 3 remain.

Stop No. 7. Switchback stop where good views of the west and north are encountered. Locate Spooner's Cove. Terrace No. 1, south of the cove has a flat surface but the same terrace north of the cove is covered by sand dunes. The dunes consist of sand blown from the Morro Bay sand spit. The Morro Bay estuary is visible inland from the spit. The valley below the viewpoint is called Reservoir Flats. It is a natural depression which the ranchers used as an irrigation pond. It is an ancient stream bed dating back to the time when terrace 2 was being eroded at sea level. Islay Creek is located a short distance to the north.

The trail makes several switchbacks before reaching the top of the hill.

Stop No. 8. This stop is at the western end of a narrow ridge which extends up the trail to the east. The ridge is the last remnant of terrace no. 5. Terrace 4 is missing from the western face of Valencia Peak. It was completely eroded during the cutting of terrace 3.

About 300,000 years ago, terrace 4 was at sea level. Terrace 5 was being cut about 450,000 years ago. Another remnant is located south of Valencia Peak. West along the ridge, the eroded edges of the folded Miguelito Shale observed at Stop No. 1 are present. Many fossils are located in the shale along this stretch of trail. Small clam shells about ¾ inches long and larger scallop shells were deposited about 6 million years ago.

Evidence exists that the narrow ridge you are walking on is an erosional remant of an ocean cut terrace. Gray sandstone similar to that found on terrace 3 and fossil borings in the shale (1/2 inch diameter) drilled by clams are present. The borings were drilled about 450,000 years ago when the rock was at sea level. Outcrops of the sandstone are present off the trail on the south side of the ridge.

At this point, the trail becomes quite steep. Shale is loose. The ridge levels out in the middle between terrace no. 5 and the old road, a possible remnant of a wave cut terrace. If this is so, terrace no. 6 could be present. Viewing down from the steep trail, the ridge top remnants of terrace 5 consists of 3 different levels. They may represent different terraces but since they are of similar elevation, they are considered to be part of terrace no.5.

The steep trail connects with an old road coming up Valencia Peak from the left. This road joins Oats Peak Trail which leads back to the Visitor's Center. Turning right leads you to the top of Valencia Peak (**Figure 23**).

Figure 23. View facing southwest of Valencia Peak from the trail leading up to the peak.

The Morro Bay sand spit which extends south from Morro Rock separates the ocean from the estuary. The alternate light and dark bands across the spit represent loose sand and vegetation respectively.

The zones of loose sand are called blowouts caused by the destruction of vegetation by human activities (**Figure 24**).

The eroded valley descending from the east edge of the park might be an ancient streambed of Islay Creek cut before repeated uplifts lifted the terraces to their current elevations.

Alan Peak is the highest pont in the park at an elevation of 1650 feet above mean sea level.

Figure 24. Morro Bay overlook showing the sand spit separating the Pacific Ocean from the Morro Bay estuary. The foreground belongs to sand dunes uplifted in Montana de Oro State Park. Morro Rock is in the background.

Part II. Geology *of Morro Bay State Park*

Geologic Setting

Morro Rock, Black Hill, Cerro Cabrillo, and Hollister Peak are the western most of 8 "morros" that extend in a straight line to San Luis Obispo. Hollister Peak is the second highest of the morros at 1400 feet. They are all erosional remnants of extinct volcanoes which last erupted about 22 million years ago. The upper part of the solidified vent as well as the sloping flanks were eroded leaving the lower part of the hardened vent standing above the surrounding softer rock. Since it solidified from molten material it is referred to as igneous rock (**Figure 25**).

Figure 25. The last of the 8 morros are shown in this aerial view, ending with Morro Rock in the foreground. Source: By Picasa posted on the internet by IPTC Photo Metadata.

The twin peaks of Pine Mountain and Rocky Butte are located along the top of the Santa Lucia Range. They are also remnants of hardened volcanic vents of the same age and composition of the morros. They are aligned with some 2 dozen smaller peaks covering a length of 20 miles.

To the left of Pine Mountain is Junipero Serra Peak at an elevation of 5800 feet, the highest elevation in the Coast Range. Cone Peak is to the left of Pine Mountain and is 5100 feet in elevation. The two peaks are 70 miles away and are visible only on clear days.

Morro Bay State Park is located in the town of Morro Bay. It is characterized by Morro Rock, the end of the 8 morro chain that erupted between 27 and 23 million years ago. In perspective, the San Andreas Fault formed around 23.5 million years ago. The morros may have formed when the East Pacific Rise approached the California Coast 25 million years ago. Subduction beneath the continental plate was occurring back then (**Figure 26 & 27**).

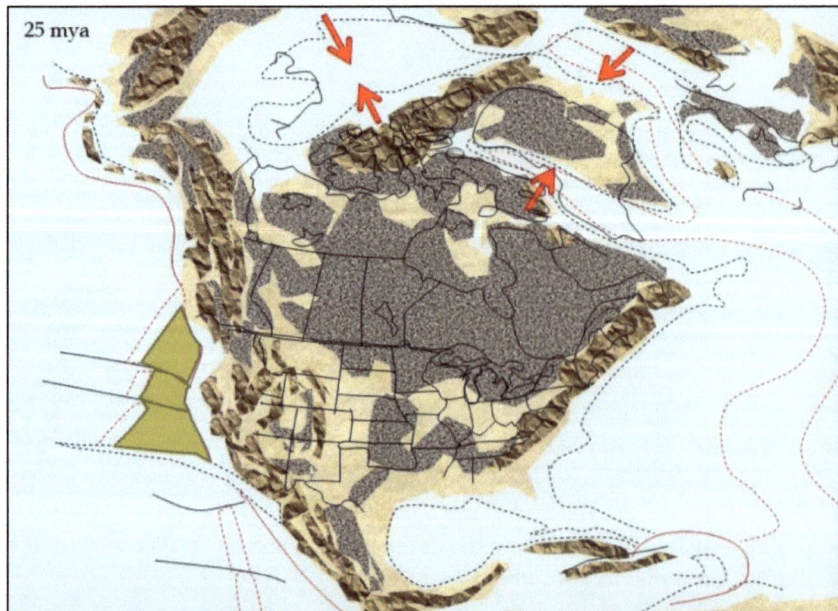

Figure 26. The East Pacific Rise (tan patch) approached the California Coast subduction zone lying off the North America Plate. As subduction occurred, molten magma rose to form the volcanic morros aligned along Highway 1. Source: Szary, 2014.

Figure 27. Close up of Morro Rock accessed from the state park. Source: Posted on the internet by vacationidea.com.

Figure 9 presents the morros on the geologic map shown in pink with symbol Toi assigned to them. The description includes Morro Rock- Islay Hill volcanic complex (listed on the map as Oligocene in age). The morros consist of porphyritic dacite, the extrusive equivalent of quartz diorite (tonalite). The composition of dacite consists of 50% phenocrysts with a typical composition of 65% andesine, 15% biotite and clay, 10% hornblende, 5% quartz, and 5% magnetite, apatite, and zeolites.

Groundmass consists of altered plagioclase, biotite, glass, quartz, and hornblende (Hall, 1973). These rocks are exposed in a series of volcanic plugs and lava domes that form distinctive peaks between Morro Bay and San Luis Obispo. Flow banding is common. Radiometric age dates are from 27 to 23 million years old (Stanley and others, 2000).

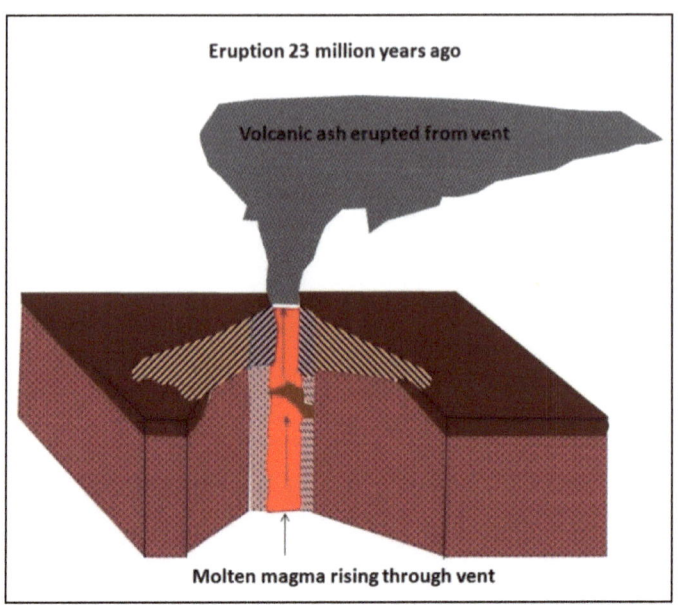

Figure 28. The eruption of Morro Rock began 23 million years ago as the East Pacific Rise approached the California coastline and was subducted beneath the continental margin. Source: uncited document posted on the internet.

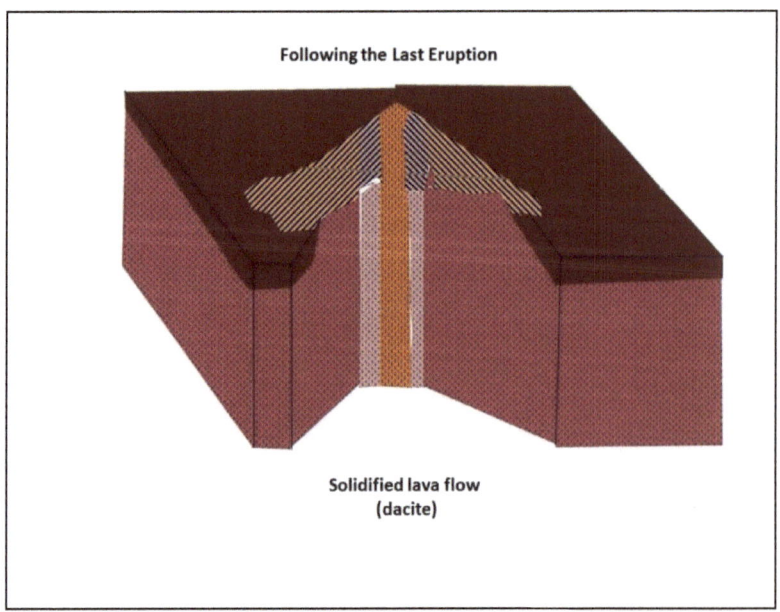

Figure 29. Morro Rock became solidified as dacite when the final eruption ceased. Source: uncited document posted on the internet.

Morro Bay State Park and the town of Morro Bay are underlain by Middle to Late Pleistocene Epoch old eolian deposits blown in by the wind from old stabilized sand dunes. They are moderately sorted, and moderately consolidated. Deposits are capped by moderately well developed pedogenic soils. They are dissected by Los Osos Creek (**Figure 30**).

The northern shoreline and sand spit consist of Holocene eolian deposits that are unconsolidated, well sorted white to brown windblown sand. Forms active sand dunes along the west side of Morro Bay (**Figure 31**).

Finally, young alluvial flood plain deposits of Holocene to Lower Pleistocene age are undivided, unconsolidated sand, silt and clay bearing on floodplains and valley floors along the shoreline leading to Morro Rock. These deposits are locally divided by relative age (2=youngest, 1=oldest).

Figure 30. Morro Bay estuary viewed from the state park. Background consists of the sand spit accessed from Montana de Oro State Park. The surface consists of pedogenic soils formed on top of old eolian deposits.

Figure 31. Uplifted sand dunes in Montana de Oro State Park in the foreground. These are the older eolian deposits described for Morro Bay. The Holocene eolian deposits belong to the sand spit in the back ground.

The End

References

Bateman, P.C., 1974. Model for the origin of Sierran granites in California Geology.

California Department of Parks & Recreation, ___. Montana de Oro State Park Detail Map.

Hall, C.A., Jr., 1973. Geology of the Arroyo Grande 15' Quadrangle. California Division of Mines & Geology.

Hummert, B., 1978. Visitors guide to Mount Diablo, California. Department of Parks and Recreation.

Stanley, R.G., Wilson, D.S., McCrory, P.A., 2000. Locations and ages of middle Tertiary volcanic centers in coastal California. OFR 2000-154. U.S. Geological Survey, Department of the Interior, Washington D.C.

Szary, W.A., 2014. Introduction to Global Plate Tectonics II: North America; Alaska, The Appalachian Mountains, The Western US, Mexico, & the South American Geologic Histories.

Wiegers, M.O., 1979. Morro Bay South 7.5 Minute Quadrangle, San Luis Obispo County, California: A digital database. California Geological Survey, Department of Conservation, California.

Book Publication Catalog
Plate Tectonic Series

All book titles are intended for those interested in earth sciences at the secondary school, community college, and first year undergraduate level of study. Technical terms are defined in the text where appropriate.

Introduction to Global Plate Tectonics I: Plate Tectonics Theory, Paleogeography, and the Ocean Basins. By William A. Szary. Earth2Energy Educational Publishing, Tampa, FL. This book is intended for those interested in general geology. Contents include many images to help the reader understand the underlying principles used to explain plate tectonics. The book summarizes plate tectonic theory by explaining the various driving forces behind continental drift theory. Text is presented in plain language for those interested in learning about the basic principles geologists use to explain the positioning of continents around the globe. Some technical terms are used, but are defined as they are presented. Chapter I **Plate Tectonics Theory** presents geologic models redrawn from the National Geographic Society, and by diagrams prepared by the United States Geological Survey. The text helps to elaborate on the image captions with more detail. Chapter II **Paleogeography** presents the principles of continental drift through global map reconstructions showing the assemblies of the continents through geologic time from the Precambrian Era Rodinia supercontinent through Gondwana, and the breakup of Pangea. The globe map series were produced by Dr. C.R. Scotese of the PaleoMap Project. Maps were used with permission from Dr. Scotese. Chapter III presents the geologic history of the major **Ocean Basins** formed by continental drift and collision throughout geologic time. A series of maps produced by the National Geographic Society, reconstruction drawings published in technical journals including the Deep Sea Drilling Project, and globe maps prepared by Robert Hall of the SE Asia Research Group are used to present graphics which accompany the narrative. **Purchase online through Createspace eStore at** 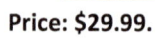**.com/4931771. Price: $29.99.**

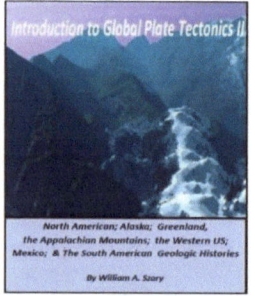

Introduction to Global Plate Tectonics II: North America, Alaska, The Appalachian Mountains, the Western US, Mexico, & South American Geologic Histories. By William A. Szary. Earth2Energy Educational Publishing, Tampa, FL. The second part of a five part series covering the subject of plate tectonics, paleogeography, and the drifting and buildout of continents. Part II covers the development of the North American basement to the present with a peek into future continental arrangements. Chapters on Alaska, the Appalachian Mountains, Western US, Mexico, and South American geologic histories are included. **Purchase online through Createspace eStore at** https://www.createspace**.com/4950697. Price: $49.99.**

Introduction to Global Plate Tectonics III: Europe, Russia, Mongolia, China, Korea, & Indochina Geologic Histories. By William A. Szary. Earth2Energy Educational Publishing, Tampa, FL. The third in a five part series covering the development of the European basement to the present day. Chapters on Russia, Mongolia, China, Korea, and Indochina geologic histories are included. **Purchase online through Createspace eStore at** https://www.createspace.com/4957439.

Price: $39.99.

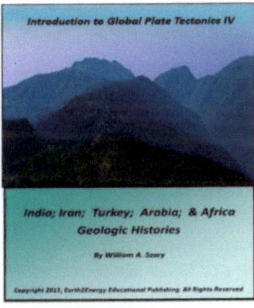

Introduction to Global Plate Tectonics IV: India, Iran, Turkey, Arabia, and Africa Geologic Histories. By William A. Szary. Earth2Energy Educational Publishing, Tampa, FL. Part four covers the development of the Indian, Iranian, Turkish, Arabian, and African basement to the present day. This book approaches the subject matter in a more technical forum.

Purchase online through Createspace eStore at https://www.createspace.com/5166539.

Price: $49.99.

Introduction to Global Plate Tectonics V: Australia & Antarctica Tectonic Evolution. By William A. Szary. Earth2Energy Educational Publishing, Tampa, FL. The last of the five part series covering the development of the Australian and Antarctica Precambrian Era basement continuing to present day. This book focuses on the more technical aspects of plate tectonic evolution.

Purchase online through Createspace eStore at https://www.createspace.com/5189356.

Price: $19.99.

Introduction to Geomorphology Series

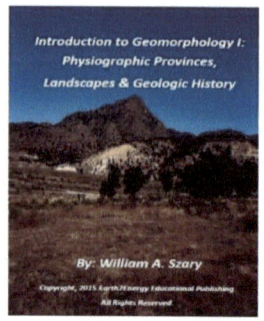

Introduction to Geomorphology I: Physiographic Provinces, Landscapes, & Geologic History. By William A. Szary. Earth2Energy Educational Publishing, Tampa, FL. Introduction to Geomorphology I reviews the geologic history behind each physiographic province providing typical and atypical photographic representations for each recognized province in the continental US.

Purchase online through Createspace eStore at https://www.createspace.com/5223816.

Price: $39.99.

Introduction to Geomorphology II: Structure, Landforms, & Geologic Processes. By William A. Szary. Earth2Energy Educational Publishing, Tampa, FL. Book II continues with expanding the description of geomorphic provinces describing landscapes in the context of geologic structure, landforms, and basic principles which shape landforms and landscapes. Many photographs are presented in this book covering constructive, destructive, mass wasting, fluvial, glacial, and coastal processes. **Purchase online through Createspace eStore at** https://www.createspace.com/5369007. **Price: $49.99.**

Florida Geology Series

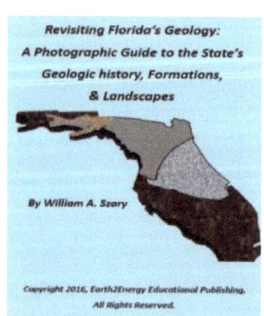

Revisiting Florida's Geology presents a summary of the state's early geologic history during the Paleozoic and Mesozoic Eras including a discussion on geologic formations, structures, and tectonic processes forming the basement complex. The Cenozoic history is presented in the context of the uppermost limestone and sedimentary rock formations. Various types of landscapes are presented using selected county geologic maps to show which formations are responsible for producing flatlands, gently to moderately sloping hills, steep hills, valleys, and karstic processes.

Purchase online through Createspace eStore at https://www.createspace.com/5931881. **Price: $59.99.**

Other Titles

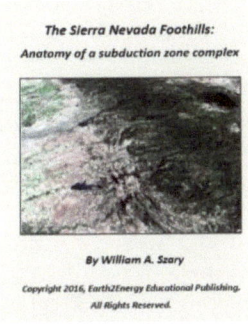

The Sierra Nevada Foothills presents an in depth analyses of the characteristics associated with a subduction zone complex based on a field study completed in the Sierra Nevada Foothills Terrane near Valley Springs, California during 2000. A proposed geologic map was compiled based on observations completed on a limited number of rock outcrop exposures, and through the use of satellite imagery obtained in 2005. A series of plate tectonic models are presented offering theories on accretion rates, erosion rates, plate collision rotation, etc. The book is intended for advanced geology students at the third to fourth level undergraduate and graduate studies level in the geological sciences. **Purchase online through Createspace eStore at** https://www.createspace.com/6231576. **Price: $19.95.**

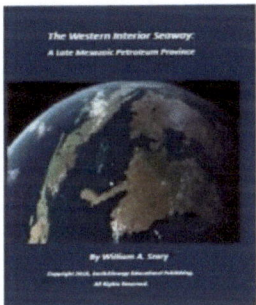

The **Western Interior Seaway** accumulated thick sediments which support hydrocarbon sources and reservoirs beginning in the Jurassic Period where the seaway was limited to the Canadian British Columbia and Saskatchewan Provinces. Seas retreated and advanced throughout the Northern Canadian Provinces up until the Early Cretaceous Period when the seaway began to invade the western U.S. states. By Late Cretaceous time, the seaway was at its maximum coverage throughout the Great Plains and along the Rocky Mountain front. The western parts were subjected to deformation while the eastern part remained undisturbed. Deformation along the Rocky Mountain front promoted fluvial floodplain and swamp deposition along the western shoreline building out deltaic deposits as the drainage reached the western shoreline. Coal deposits accumulated. The eastern shoreline consisted mostly of terrestrial deposits reworked as they invaded the seaway. Marine limestone accumulated in the quieter waters while sandstone, shale, and mudstone buried limestone under similar conditions. These sediments promoted hydrocarbon development. **Chapter 1** provides an overview of the Western U.S. Mesozoic and Cenozoic geologic history. **Chapter 2** describes the stratigraphy and sedimentation associated with the seaway during transgressive-regressive cyclic deposition. **Chapter 3** presents coal and petroleum resources of Canada tied to the Western Canadian Sedimentary Basin. **Chapter 4** presents petroleum provinces in the Western U.S. developed by the Western Interior Seaway. **Chapter 5** summarizes the paleogeography, and hydrocarbon development. Price: $49.99. **Purchase online through Createspace eStore at** https://www.createspace.com/7836912.

The Cascadian Foothills: Shifting geosynclines, depositional environments, & tectonic rotation is about the Lewis County Cascadian Foothills in southwestern Washington which represents a diverse geologic history of marine transgressive-regressive depositional cycles, shifting geosyncline troughs, and a chaotic assemblage of faults and folds exposed to convergent margin plate subduction and plate rotations occurring throughout the Tertiary Period. Field mapping techniques and satellite imagery were used to reconcile confusing attitudinal observations as well as primary and secondary folds observed by superimposition. Chapter 1 presents regional plate tectonic setting, stratigraphy, Early Eocene tectonic history, a brief summary of the North Cascade sub continental collision, and an overview of the western geosyncline basin shifting occurring during Middle Eocene through Miocene Epochs. Chapter 2 presents study area stratigraphy observed in central Lewis County for the McIntosh, Northcraft, Skookumchuck, and Lincoln Creek formations. Depositional models for each formation are presented. The Northcraft formation was subdivided into two areas: the West Central Volcanics and the East Central Volcanics based on the presence of andesite flows and pyroclastic ash deposits each considered mappable units in central Lewis County. The Skookumchuck Formation exhibited cyclothemic properties based on observed exposures. Two models are offered based on ideal cyclothem models derived from the Pennsylvanian Period for comparison with Middle to Upper Eocene cyclothemic deposition. Chapter 3 provides an overview of geologic structure and tectonic rotation with impacts on regional deformation patterns related to folding and faulting. Plate rotation theories are presented along with southwest Washington regional studies and local study area rotational models. Plate rotation geochronologic estimates are provided for the Middle Eocene, Oligocene, and Miocene timelines. Geosyncline influences and study area tectonic sedimentation are discussed.

Should the purchaser have an interest in reviewing sample chapters of any title, please direct inquiries to the email address (wszary@netzero.net) with the requested title. A PDF version will be provided for the first Chapter free of charge. To order any title, the Purchaser may copy the catalog, circle the title, and include with remittance to the following address or visit the publisher's eStore web page at the address shown for each title for direct purchasing. **Price: 45.99. Available through Kindle Direct Publishing.**

Earth2Energy Educational Publishing
Port Richey FL 34668

www.ingramcontent.com/pod-product-compliance
Lightning Source LLC
Chambersburg PA
CBHW040253220526
45473CB00001B/464